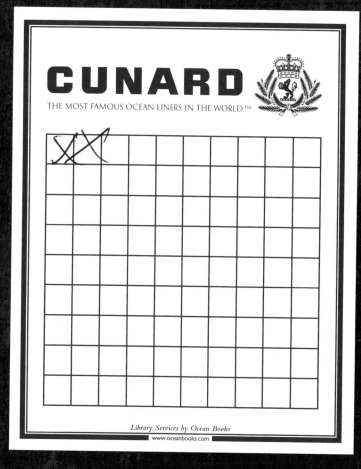

CUNARD

THE MOST FAMOUS OCEAN LINERS IN THE WORLD ™

DARLYNE MURAWSKI | HAWAII | *A carnivorous caterpillar uses its grappling limbs to seize a fruit fly in a tight embrace.*

DEADLY INSTINCT

MELISSA FARRIS

Foreword by
BRADY BARR

NATIONAL GEOGRAPHIC

WASHINGTON, D.C.

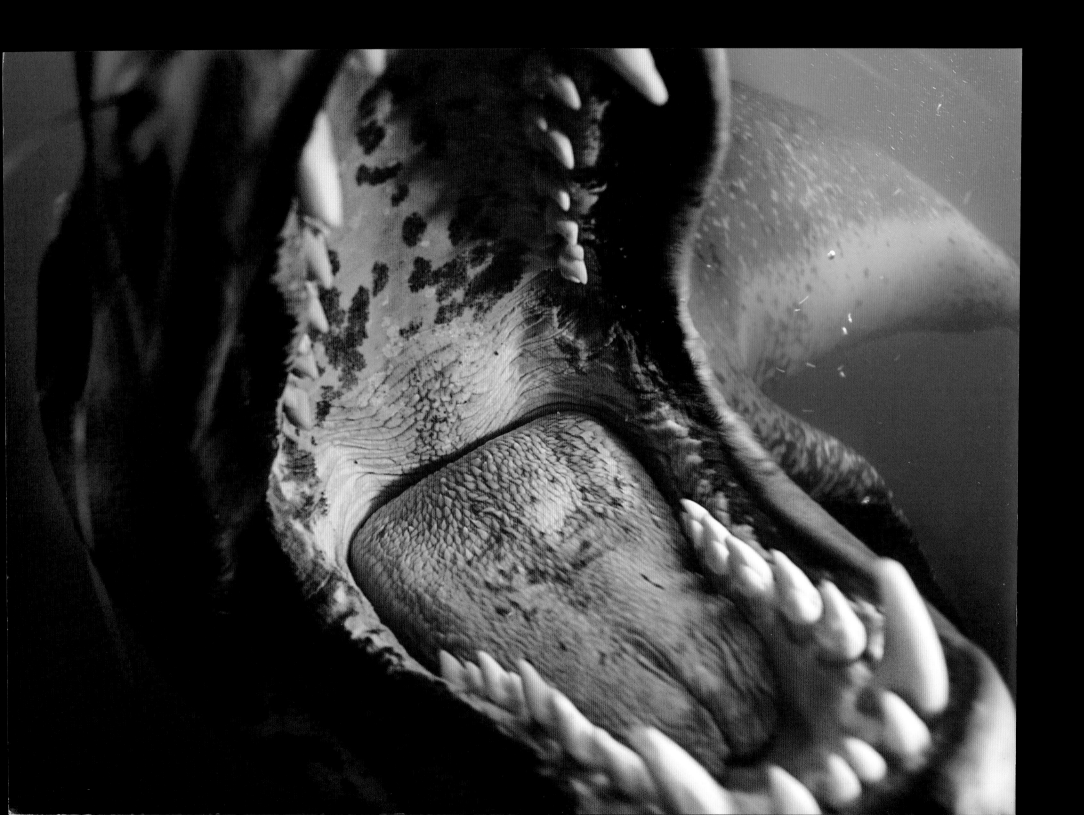

{ CONTENTS }

PAGE 1: MIKE PARRY | SOUTH AFRICA | *Bursting from the waves, a great white shark snatches a seal decoy in a viselike grip.*
OPPOSITE: PAUL NICKLEN | ANTARCTICA | *A leopard seal flashes a threatening set of teeth.*

CHRIS JOHNS | BOTSWANA | *Roughhousing among pack members is an everyday activity for African wild dogs.*

DAN STAHLER | WYOMING | *Zeroing in on their target, a pack of wolves surround a bison cow they have isolated from the herd on a frozen creek in Yellowstone National Park.*

BEVERLY JOUBERT | BOTSWANA | *African lionesses, the primary hunters in a pride, down— and start to devour—an African buffalo.*

OPPOSITE: TANYA MANFREDIZ | FLORIDA | *Herpetologist Brady Barr cautiously approaches an alligator while making an episode of* Deadly Encounters *on location in the Everglades.*

FOREWORD BY BRADY BARR

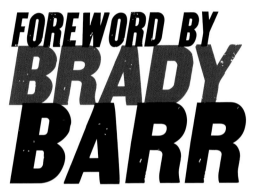

Reptile expert Dr. Brady Barr has made it his life's work to study and protect some of the world's most dangerous and endangered land animals—alligators and crocodiles. He is the first person to capture and study all 23 species of wild crocodilians, and is the host of Dangerous Encounters *on Nat Geo WILD.*

I have worked with wild animals for over 20 years, 13 of them as a resident scientist for National Geographic. As a kid, I was fascinated by our natural world. I grew up in Indiana, where cornfields stretched as far as the eye could see and about the wildest thing a kid could come across might be a squirrel. I got my introduction to the more exotic species on the planet at zoos and aquariums, which fueled my interest and passion. As I grew older and had more in-depth outdoor experiences, I realized that the zoo animals I so loved were missing something. They had lost their wildness— that unpredictable, unbridled, survive-at-any-cost mind-set. It's the quality that's so supremely special about our natural world. Don't get me wrong, the animals in zoos are fantastic. I am a product of American zoos, where the animals are ambassadors for their species in the wild and play crucial roles in generating interest and passion in kids all over the world. But zoo animals by no means accurately represent the majestic nature of being "wild." An insatiable appetite for interacting with wild animals in their natural habitats has driven my career. It doesn't matter who you are, or how much experience you might have; wild animals simply never cease to surprise and amaze.

Not long ago, I was on the Nile River in Uganda. I had just captured a croc for a research project, one of 5,000 or so I have caught over the years, and was preparing to release it. Releasing a croc involves untying its legs and snout and then jumping off its back. The animal either runs or walks back into the water, with the occasional snap of the jaws as a reminder of who is really boss. I have done this thousands of times, and the procedure nearly always follows this script, but this time the croc had other ideas. The scaly old croc raised itself off the ground and slowly walked toward the water; then, instead of gliding away in the current, it abruptly stopped and turned toward my inflatable boat, which was pulled up on the riverbank. He placed first one foot, then a second on the dinghy's side, then pulled his massive body up and into the tiny raft, where he lay down and closed his eyes for a nap. I was speechless. If that croc punctured my boat, I'd be trapped along a remote part of the river. After a long, long, long rest, the old croc lumbered up and out of the boat and back into the water. Cue a huge sigh of relief. I could fill a book with such experiences.

During my career I have worked in more than 70 countries and in some of the most remote places on the planet; I've dealt with predators from giant squid to giant snakes, from Komodo dragons to electric eels, and all critters in between. The most important thing I have learned is that being wild is special. It means being unpredictable, astonishing, awe-inspiring, and dangerous. Being *WILD* is the real deal. Wildness has almost cost me my life on several occasions, but it also keeps me coming back to nature. It is the fuel that feeds my passion. I have an insatiable desire to learn more, see more, and experience more of the quality of being wild, something I feel this book captures perfectly . . . and, maybe most important, it allows you to taste wildness from the safety of your couch!

HUNTING WITH A CAMERA

"[I]t is not necessary to convert the wilderness into an untenanted and silent waste in order to enjoy the sport of successfully hunting wild birds and animals." –George Shiras III

More than 100 years ago, *National Geographic* magazine unveiled its first issue devoted to wildlife photography. Published in July 1906, the edition featured more than 70 black-and-white images of animals in the wild. Several had been shot at night by the pioneering photographer George Shiras III as he paddled along silent waterways, his camera mounted on the prow of a canoe, illuminating his subjects with the startling flash of ignited magnesium powder. Page after page of the magazine showed moose and deer feeding at the water's edge, caught on film before they could even react. Shiras was a visionary in trip-wire photography as well, which he used to freeze raccoons, beavers, and porcupines forever in time; they had taken their own pictures, so to speak, by tripping a line fastened to the shutter of a camera placed nearby.

Brought to the public eye after more than a decade of experimentation and innovation by Shiras, these images were some of the first to show the everyday behavior of wild animals in their natural habitat. Alongside these rarities ran an appeal from Shiras for sportsmen to put down their rifles and pick up cameras instead.

Shiras had spent much of his youth hunting and fishing on Michigan's Upper Peninsula. Though he left the area for college, law school, and a career in politics, the beauty of those untamed forests made a lasting impression on him, so much so that he returned nearly every summer for the next 70 years. As a congressman representing Pennsylvania's 29th District, Shiras introduced the first federal legislation that protected migrating birds. His work on the Public Lands Committee helped set aside vast areas for conservation in what ultimately became Olympic National Park, Petrified Forest National Park, and an extension of Yellowstone National Park, where he discovered a species of moose that was eventually named for him.

Shiras did not portray animals as they were customarily depicted by photographers of his time—stuffed and mounted on a wall, or lying dead beneath the foot of a triumphant hunter. Rather, he captured them as they truly existed in nature—alert and sentient beings, active and reactive, filled with instinct and intelligence. His greatest legacy may therefore be the reverence for wildlife that his images brought to the world.

The conservation-minded work that began with George Shiras III remains vibrant today. Wildlife photography has become a highly specialized field, but despite a multitude of technological advances, the innovation and perseverance of early practitioners like Shiras are still highly prized—and much in demand. Indeed, the pages that follow showcase some of the most arresting work by modern lensmasters, each of whom has developed his or her own unique perspective on the deadly instincts that allow animals to survive in the wild.

Take Darlyne Murawski: Inspired as a child by the bugs, birds, and plants she found in her backyard—and by the close-up photos she came across in the pages of *National Geographic*s from the 1960s—Murawski studied photography in art school. From there she went on to earn a Ph.D. in biology at the University of Texas, eventually studying butterflies in the rain forests of Costa Rica. There Murawski got hooked on shooting macro portraits of insects and flowers, and realized that her camera was a powerful tool for expressing both her art and her fascination with the natural world. "The camera took me outdoors," she says, "and let me observe the creatures I was interested in."

Murawski credits her fly-high views of insect life—as well as her portraits of birds, of other animals, and of their habitats—to a marriage of technique and temperament. "I guess I have intrinsic patience," she says. "And that's a good thing, because if you want to get a good picture, you have to be persistent. My art background has helped me to see details—and to see ways of portraying things—while my biology background has helped me learn about the creatures themselves."

Chris Johns, too, knew a thing or two about the wild lives of wildlife before he became a contract photographer for *National Geographic* in 1985. After helping to raise animals on the small farm in Oregon where he grew up, Johns was preparing to become a large-animal vet when he happened to take an elective class in journalism at Oregon State University and got bitten by the shutter bug. He went on to work as a staff photographer at a couple of big-city dailies before landing at the *Geographic*.

An early assignment from the magazine took Johns to Africa's Great Rift Valley, whose animal and human inhabitants enthralled him. He also swiftly perceived the fragility of their symbiosis. "I saw how quickly the world was changing, and the profound impact this was having on wild animals and their habitat, as well as on people and their culture," reports Johns, now the Editor in Chief of *National Geographic* magazine. "At that point I became committed to documenting the changes under way, the forces responsible for those changes, and what it all meant in people's lives—and in the lives of these animals."

The photographer Joel Sartore likewise sees wildlife images as potential catalysts for change. Covering America's coastlines on a magazine assignment in the early 1990s, Sartore was horrified to discover the beaches of Galveston, Texas, blackened by tar and littered with garbage and medical waste that had washed ashore. "I found a dead dolphin wrapped up in plastic ropes," he says. "In one place after the other on the Gulf Coast, I found the environment had just been trashed."

From that day on, Sartore resolved that his work would help preserve the natural world. Though he is still an accomplished field photographer, Sartore devotes much of his time these days to shooting stark studio portraits of animals, many of which appear on the endangered species list. The goal of these more intimate views of wild animals is to forge a personal connection between observer and subject. "I am interested in using photography to right wrongs, socially and environmentally," says Sartore. "I think of these pictures as emissaries, and they can get to work."

The images in this book—our two-dimensional ambassadors from the wild—are drawn from the nearly 11 million images housed in the National Geographic Picture Collection. Our hope is that they will paint a vivid picture of the grueling challenges that life on Earth throws at members of the animal kingdom. Even in the absence of human-induced habitat destruction, wild animals face lethal threats from one another on a daily basis. Many of them wage fierce battles designed to let them feed, breed, grow, and carve out territory for themselves. Precipitated by nothing more than the animals' natural survival instincts, such deadly encounters form a leitmotif of this book.

View—if you can stomach it—the chapter on predation, in which photographers capture, time and again, the precise moment a life-or-death battle between predator and prey is decided. Following that is a chapter showcasing the fascinating survival mechanisms some animals have evolved. Many of these ruses provide scant seconds of diversion—but in the wild, a few heartbeats often mean the difference between escape and demise.

For all animals, survival also means propagating the species. For some animals, as chapter 3 makes clear, the arrival of breeding season triggers dramatic, ritualized battles that determine which members of a group will prevail in passing down their genes. The final chapter turns from the turmoil of existence to the pleasures of play, showing how and why roughhousing can be educational: Along with chasing, running, climbing, and playing with objects from their environment, rough-and-tumble play-fighting helps the young of a species develop essential motor skills and cognitive abilities. Play may also help forge and reinforce social bonds, and continues for some animals well into adulthood.

If the images collected in the pages that follow can serve a larger purpose, it may be to reflect our own nature, behavior, and experience: The struggle to provide food, shelter, and safety for our families and ourselves; the canny ways in which we adapt to sidestep harm; the fiercely competitive drive to win favor with a mate, or to win primacy in a group; the inquisitive, creative compulsion to explore our surroundings and make connections with our neighbors—all of these attributes we "civilized" animals share with our "wild" counterparts.

Looking through a camera lens, it turns out, is often much like gazing into a mirror.

ES OF FEAR

PREDATOR PARASITE & PREY

Death is a constant, looming presence in the wild. It creates an environment of stress and fear that can shape the eating patterns and breeding habits of prey animals—and sometimes the very ground beneath their feet.

In the winter of 1995, the gray wolf was reintroduced in Yellowstone National Park to an elk population that had lived free of lupine predation for more than 50 years. Within five years, the behavior of the elk began to change. They adjusted their grazing patterns and kept on the move. They summered at higher elevations, possibly to avoid areas where wolves denned with pups. They gathered into larger groups while feeding in wintertime, seeking safety in numbers. They fed less in open, grassy areas—where wolf kills had become common—when food was plentiful elsewhere in the park. In selecting feeding ranges after the wolf reintroduction, the elk traded some favorite foraging ground for a reduced risk of attack. Such areas of predator influence—some animal behaviorists call them "landscapes of fear"—demonstrate the power a predator can exert over the behavior of its prey. It's not just a question of kill or be killed in the natural world; sometimes it's a choice between eating or being eaten.

Not every creature that feeds upon another is a predator, however. Consider the vampire bat: This mammalian parasite bites a victim, drinks its blood, then disappears into the night, leaving its quarry alive to dine upon another day. Some parasites, such as the howler monkey botfly, can live nearly half their lives inside a host body, sapping its strength and resources but ultimately leaving the host alive. Less charitable are brood parasites such as the common cuckoo, which tricks other bird species into rearing its young. First, a cuckoo destroys an egg in another bird's nest and replaces it with one of its own. After hatching, the cuckoo chick systematically pushes its rival eggs (or even hatchlings) from the nest, killing them. Meanwhile, the unsuspecting host parents—helpless to deny their instinctive responses to the cuckoo chick's screeches for food—feed and nurture the murderous impostor until it outgrows the nest.

The true power brokers of the animal kingdom are the apex predators. These mighty hunters stand atop a food chain, with virtually no predators of their own. The influence they wield through a real or merely imagined attack can shape their environment in ways that wildlife biologists are still studying, including in Yellowstone. During the park's half-century-long wolf drought, the elk devoured so many streamside willows and cottonwoods that the soil along the banks eroded, altering the floodplain of the Gallatin River. Since the wolf's reintroduction, those trees have begun to rebound, and some researchers link the recovery to the wolves' presence. Reinforced by root systems and sheltered by ungrazed trees, the stream banks now furnish increased habitat for beavers, birds, fish, and insects. The park's coyote population has declined, providing a safer haven for the small mammals, such as pronghorn fawns, that make up their prey. And the elk herd, by increasing its "vigilance"—the time spent watching, listening, and smelling for predators—has begun to calculate its routines with the appetite of its rapacious new neighbors in mind. As the elk have learned, there are many ways to die within the landscape of fear, but only one way to live—carefully.

CHRIS JOHNS | AFRICA | *These images show a method of cheetah hunting rarely documented, as a duo of African cheetahs force an exhausted antelope underwater, holding it there until it drowns.*

REVONDA GENTRY | OHIO | *Catapult-like muscles in the tongue of this captive chameleon enable it to lash out at more than 13 miles an hour, snatching a cricket in midair.*

OPPOSITE: BRUCE DALE | MEXICO | *A vampire bat drinks blood from a sleeping calf. The bat requires only about a tablespoon of blood a day to sustain itself.*

Part of the secret of success in life is to eat what you like and let the food fight it out

INSIDE.

–MARK TWAIN

OPPOSITE: BIANCA LAVIES | NEW YORK | *A yellow timber rattlesnake begins to consume a mouse, a meal that could take as long as four days for the snake to digest.*

OPPOSITE: JOEL SARTORE | BRAZIL | *A group of caimans wait for unwitting fish to swim downstream.*

TUI DE ROY | GALÁPAGOS ISLANDS | *Open wide: A brown pelican dives underwater, scooping up a fish in its beak's elastic pouch.*

WILD ENCOUNTERS

Photographer CHRIS JOHNS on capturing a battle between deadly foes:

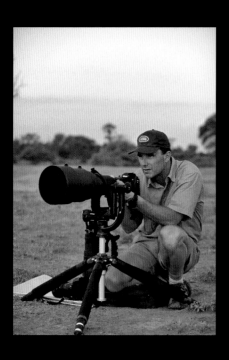

Chris Johns studied to be a vet at Oregon State University, then got hooked on photography and wound up shooting football and other sports for the Topeka Capital-Journal. The skills and reflexes he gained from those experiences came in handy in Tanzania in 1988, when Johns photographed a deadly encounter at dawn.

"A pride of lions were tearing at an African buffalo and jumping on his back," he recalls, "and he was bellowing because they were essentially starting to eat him alive. They'd pull him down but he would struggle to his feet and make another stand, swatting at his attackers.

"I felt empathy for the buffalo," says Johns, now the Editor in Chief of National Geographic magazine. "But I identified with the lions too—these episodes are not without peril for the predators. I was privileged to witness the run-in, and I took pains not to affect its outcome. I never want to be an agent of the action; I just want to capture it, and through those pictures evoke an appreciation of the magnificence of these creatures."

ABOVE: COURTESY OF CHRIS JOHNS | AFRICA | *National Geographic photographer Chris Johns sets up his equipment for a shot while on assignment in Africa.*

OPPOSITE: CHRIS JOHNS | TANZANIA | *Two lions momentarily get the upper paw in a deadly battle with an African buffalo. The contest continued for more than two hours before the prey succumbed.*

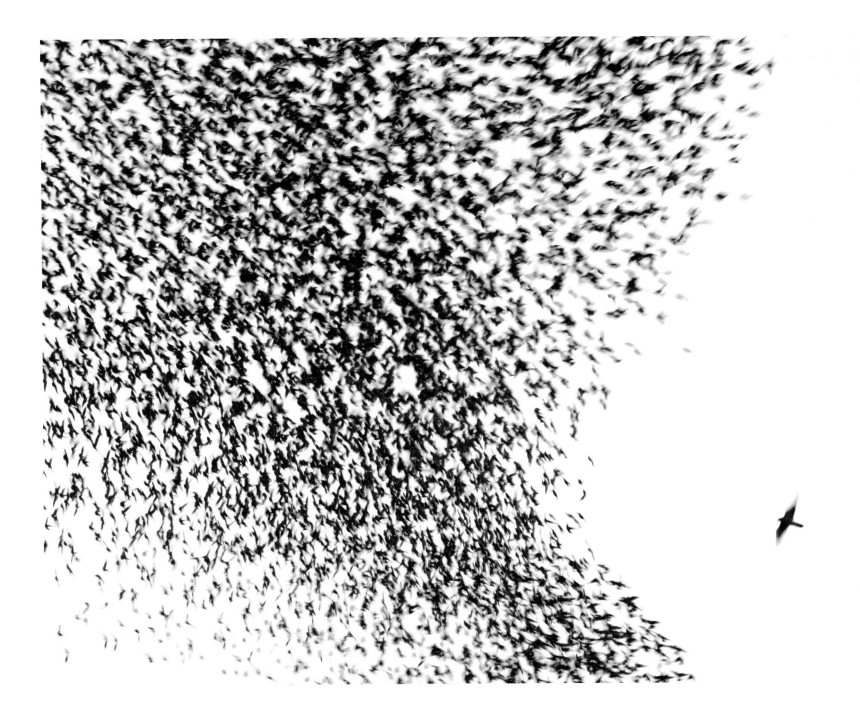

PAGE 34: FRED BAVENDAM | SOLOMON ISLANDS | *Like iron filings repelled by a magnet,*
schooling fish flee a gray reef shark on the hunt.

PAGE 35: MANUEL PRESTI | ROME | *A flock of starlings maneuver as one to escape the strike of a peregrine falcon.*

RANDY OLSON | RUSSIA | *A brown bear goes fishing—successfully—for spawning Pacific salmon in Kamchatka's*
Kurilskoye Lake. Kamchatka salmon are threatened by overfishing and caviar poaching, and that in turn imperils
the bears and other animals that rely on the salmon for sustenance.

JOEL SARTORE | YELLOWSTONE NATIONAL PARK | *Once the nation's most widespread carnivore, a reintroduced gray wolf pursues a pack of scuttling elk. Though the gray wolf is a skilled hunter, there's no guarantee of victory in this enterprise: Some Yellowstone wolves have been killed by elk.*

STILL-HUNTING

is the polar bear's primary means of predation.
Upon finding an active seal breathing hole in the ice,
the bear lies motionless and silent on its stomach,
nose to the hole, waiting for its target to surface.
Once a seal emerges for a breath, the bear dispatches
it with a quick bite—or even a blow to the skull.

OPPOSITE: JOEL SARTORE | ALASKA | *Top predators and fearsome hunters, polar bears also scavenge food when it's available. Here a polar bear dines on a found meal of a bowhead whale carcass.*

PAGE 42: CHRIS JOHNS | HAWAII | *A Culex mosquito perches near the eye of an 'i'iwi, a bird native to Hawaii. The insects have spread deadly avian malaria across the state, devastating Hawaii's bird populations.*

PAGE 43: JOE PETERSBURGER | ROMANIA | *Even plants prey: Here an insect-eating plant known as a great sundew has snared a damselfly in its adhesive grip.*

MICHAEL NICHOLS | INDIA | *A tigress makes good use of her stripes, taking cover while hunting in the tall grass (opposite); above, her cub feeds on one of her kills.*

PAGE 46: JOE PETERSBURGER | PANAMA | *A praying mantis strikes faster than the blink of a human eye, swiftly sealing the fate of this dragonfly.*

PAGE 47: MARK MOFFETT | ARIZONA | *A crab spider practices cannibalism, dining on one of its own kind.*

MITSUAKI IWAGO | TANZANIA | *Two spotted hyenas chase a blue wildebeest mother and calf. Though infamous for scavenging, spotted hyenas are in fact formidable pack hunters.*

I want my food

DEAD.

Not sick, not wounded: dead.

–WOODY ALLEN

OPPOSITE: JOEL SARTORE | MINNESOTA | *A captive gray wolf snarls over its bounty, the carcass of a deer.*
You wouldn't invite this one indoors, but the gray wolf is the ancestor from which all domesticated canine breeds descend.

PAUL NICKLEN | ANTARCTICA | *A female leopard seal hunts down a penguin, then offers her kill to the photographer before diving to consume her prey. Leopard seals grow to more than ten feet long, making them the most powerful hunters of all seals—and one of Antarctica's chief predators.*

PAGE 54: CHRIS JOHNS | AFRICA | *Cheetahs' teeth are too short to kill large prey like this impala, so they suffocate bigger victims by clamping down on the windpipe with powerful jaws.*

PAGE 55: CHRIS JOHNS | HAWAII | *Brought to Hawaii in 1883 in an attempt to control rat infestations in sugarcane fields, the mongoose now habitually preys on ground-nesting birds such as this Hawaiian goose.*

OPPOSITE: CHRIS JOHNS | BOTSWANA | *Spattered with blood, an African wild dog pauses from feasting on his kill to scan for members of his pack, who join in on the meal after the hunt leader enjoys the privilege of dining first.*

ABOVE: CHRIS JOHNS | BOTSWANA | *Unusual prey for wild dogs, this unlucky warthog was first the target of playful pursuit by pack youngsters. Once the pack adults took notice, the warthog was dead in less than 60 seconds.*

BIANCA LAVIES | MARYLAND | *The camera freezes a green tree frog mid-leap, its sticky tongue adhering to a damselfly.*

OPPOSITE: CHRISTIAN ZIEGLER | BARRO COLORADO ISLAND | *This mantled howler monkey is infested with a species of botfly that parasitizes howlers exclusively. Warbles on its neck and abdomen show the airholes through which the botfly larvae breathe.*

The whole of nature,
as has been said, is
a conjugation of the verb

TO EAT,

in the active and in the passive.

–WILLIAM RALPH INGE

OPPOSITE: BRUCE DALE | CONNECTICUT | *Undaunted by the mismatch in size, a lizard given to a captive (and uninterested) spear-nosed bat as food bites the nose of its would-be predator.*

GEORGE GRALL | FLORIDA | *A snapping turtle uses its cryptic coloring to bring prey within seizing range.*
The turtles are voracious, and will eat plants, fish, and fellow bottom-feeders such as crayfish.

OPPOSITE: BILL CURTSINGER | HAWAII | *The first flight of this luckless albatross fledgling was its last: It fell prey to the onrushing tiger shark.*

JOHN WATKINS | EUROPE | *A common cuckoo hatchling pushes the eggs of its host parents from the nest, terminating its competition for food.*

GOING HEAD TO HEAD

In the competition for carrion in the wild, the Rüppell's griffon vulture has a head built to keep it in the game. Its sharp beak enables the bird to slice quickly through viscera and muscles. The vulture's head and neck—both long, slender, and featherless—are perfectly adapted for plunging into dead animal carcasses to quickly snare gobbets of meat. Topping off its offal armaments is the griffon vulture's tongue: It is lined with backward-facing barbs that help tear flesh from bone.

PAGE 68: YVA MOMATIUK & JOHN EASTCOTT | KENYA | *A herd of blue wildebeests step lively to escape a crocodile during a crossing of the Mara River.*

PAGE 69: MICHAEL & PATRICIA FOGDEN | COSTA RICA | *An eyelash viper gold morph launches a strike at a rufous-tailed hummingbird.*

OPPOSITE: SUZI ESZTERHAS | TANZANIA | *Rüppell's griffon vultures and white-backed vultures share the prize of a scavenged zebra carcass.*

THE STRUG

CHAPTER TWO

GLE FOR LIFE

OPPOSITE: PIOTR NASKRECKI | SOUTH AFRICA | *When threatened, the koppie foam grasshopper exudes toxins as a defensive behavior.*

STRATEGIC SURVIVAL TACTICS

A ceaseless struggle takes place in the battlefield that is the natural kingdom, driven by conflicting motives: On one hand is the need to feed, on the other the desire to live. Adaptation is essential to achieving both goals.

In the seemingly tranquil lakes of Oregon's Willamette Valley lurks one of the world's deadliest creatures. Approximately five inches long, brown with an orange underbelly, it teems with enough neurotoxin to kill 25,000 mice, several adult humans, or virtually any of its potential predators. It is *Taricha granulosa,* the rough-skin newt, and it is locked in what is known as an evolutionary arms race with its sole toxin-resistant predator—the common garter snake.

Most rough-skin newts produce this deadly tetrodotoxin, but in the 1990s, scientists discovered that the toxin levels of Willamette Valley newts far outstrip those of newts found elsewhere in the Pacific Northwest. Garter snakes in this area have developed special survival countermeasures of their own, displaying much higher resistance to the neurotoxin than do nearby populations that live outside the newt's range. Over countless millennia and untold lineages of both snake and newt, the Willamette Valley has served as a crucible of counteradaptation, with each species constantly striving to best the defenses evolved by the other: The more toxin was produced by the newts, the more poison resistant the snakes were forced to become.

Toxic deterrents like these are usually well advertised. Warning signs—or aposematic signals, in biology vernacular—frequently take the form of bright or contrasting color patterns. The orange belly of the rough-skin newt; the bright blue, green, yellow, and red patterns of poison-dart frogs; the yellow and black stripes sported by many stinging insects—all of these broadcast that the prey represents not so much a toothsome meal as a treacherous one.

At the opposite end of the spectrum, many animals avoid becoming prey by eluding predator detection. To do this, they often employ cunning forms of deception. The dead leaf mimetica, a katydid that strongly resembles a discolored tree leaf—complete with leaf veins, browned edges, and mock insect chewspots—blends seamlessly into its arboreal background. Similarly, the early larval stages of the tiger swallowtail caterpillar resemble decidedly unappetizing bird droppings. But the caterpillar's disappearing act doesn't stop there; in its final stage it turns a bright green, developing false eyespots on its tail that lend it the menacing appearance of a snake's head.

Helpful though they may be in one aspect of the fight to survive, adaptations can come at a cost. Evolutionary changes such as developing a resistance to toxic prey can have negative effects on other biological functions. Indeed, those garter snakes with the highest immunity to newt toxins are measurably slower than snakes with lower resistance. This reduction in speed makes the garter snake more vulnerable to predators, and is thought to be a long-term consequence of the inherited resistance itself. But in an intriguing turn of events, researchers have posited that the garter snake may be gaining an unexpected benefit by consuming the toxic newt: Because the neurotoxin takes several weeks to dissipate in the snake's body after it eats a newt, the snake itself may become poisonous prey to its own predators during that time.

FROM LEFT: PEYTON HALE, MARK MOFFETT, PAUL ZAHL | *Poison-dart frogs are named for the alkaloid toxins they secrete. The toxin of the golden poison frog (above, right) is applied to the tips of darts used by the Emberá Chocó people in Colombia to bring down game.*

RAYMOND MENDEZ | TEXAS | *When menaced by a predator, the Texas horned lizard will sometimes shoot a stream of blood from its eye into the face of the troublemaker.*

OPPOSITE: BIANCA LAVIES | UNITED STATES | *Fending off a perceived threat, a northern puffer fish swells up like a balloon by inflating a belly sac with water or air. This survival stratagem makes the fish too large for predators to swallow.*

FALSE ADVERTISING

The insect on the left is a hoverfly, stingless and tasty to birds. The insect on the right is a yellow jacket, capable of inflicting multiple stings on a predator. Belying its lack of defenses, the hoverfly has evolved to bear the same warning coloration as the wasp. This adaptive strategy is known as Batesian mimicry; its payoff for the hoverfly is a reduced risk of attack.

OPPOSITE: ROBERT SISSON | UNITED STATES | *Harmless hoverflies like the one on the left sometimes wave their forelegs over their heads to more closely mimic the longer antennae of the insects they imitate.*

PAGE 82: CHRISTIAN ZIEGLER | PANAMA | *Two mimetic katydids with stem-shaped legs— can you see them?—use deception to elude detection.*

PAGE 83: JOEL SARTORE | EQUATORIAL GUINEA | *In a practice known as autotomy, geckos sometimes leave their tails behind to escape from predators.*

MITSUAKI IWAGO | TANZANIA | *African buffalo, shown here in a defensive formation, greatly outweigh lions.*
They have been known to kill the big cats to protect their herd from predation.

WILD ENCOUNTERS

BRIAN SKERRY on the legendary defenses of the jumbo squid:

Brian Skerry was diving among Humboldt squid (aka jumbo squid) at night in the Sea of Cortés when he decided to learn more about their nasty reputation: Why did local fishermen call the squid diablo rojo, *red devil*?

So he called his friend Roger Hanlon, a cephalopod expert at the Marine Biological Laboratory in Woods Hole, Massachusetts.

"Hey, Roger," Skerry told him, "I'm down here in Baja Mexico—what you can tell me about Humboldt squid?"

"I can tell you not to get in the water with them," said Hanlon. "One of them could swim away with you and me both."

"Uh, actually, Roger," Skerry replied, "I was in the water with about 40 of them last night!"

"That's crazy!" Hanlon told him. "They'll grab almost anything in the ocean that moves!"

Skerry got a taste of their ferocity the night he shot the photo opposite. "A couple of squid tried to grab my camera housing and wrench it away. But I turned the tables: I charged one and **it turned bright red—I know anthropomorphism is frowned upon, but it looked almost angry.** Then it shot out that gossamer ink cloud you see in the image."

ABOVE: MAURICIO HANDLER | JAPAN | *A portrait of photographer Brian Skerry on location.*
OPPOSITE: BRIAN SKERRY | ATLANTIC OCEAN | *The jumbo squid's color-changing ability is thought to be a mode of both communication and camouflage.*

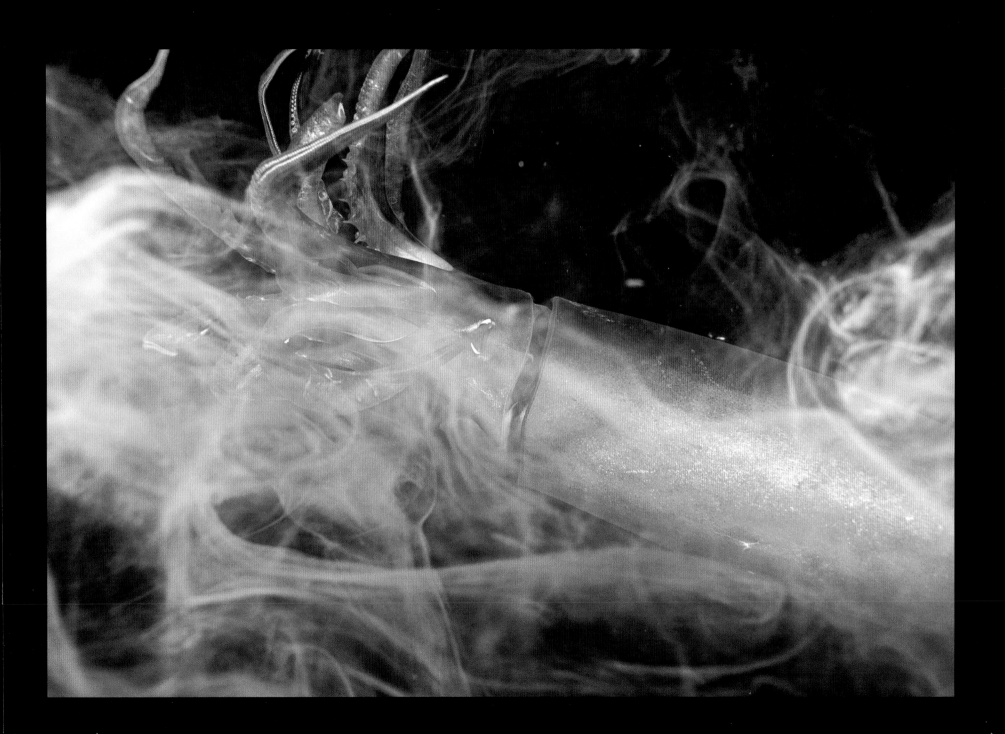

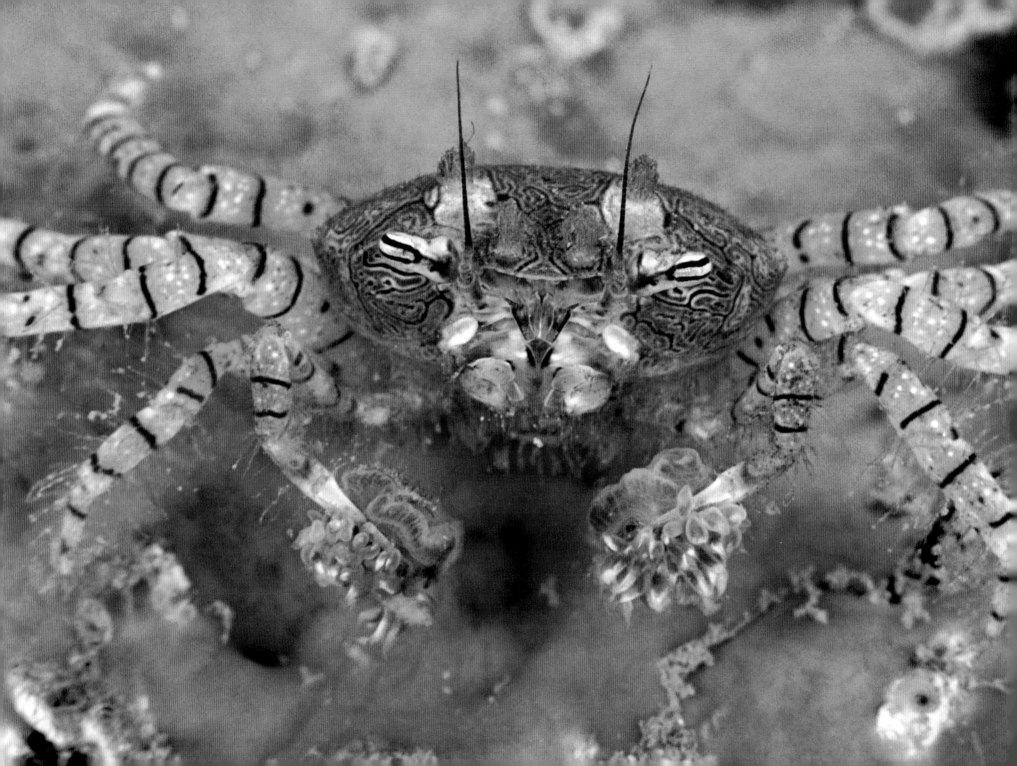

The rabbit runs faster than the fox,
because the rabbit is running

FOR ITS LIFE,

while the fox is only running for his dinner.

–RICHARD DAWKINS, AFTER AESOP

PAGE 90: GEORGE GRALL | PERU | *A white-lined tree frog adopts a defensive posture: It curls up and plays dead.*

PAGE 91: JOEL SARTORE | NEBRASKA | *When threatened by a predator, the southern three-banded armadillo rolls itself into an armored ball that shields its head, legs, and tail.*

OPPOSITE: KONRAD WOTHE | AUSTRIA | *Fabled for its speed, a European hare can run from a predator at speeds up to 45 miles an hour. It follows an unpredictable, zigzag path as a way to throw off pursuers.*

PAGES 94–95: ANUP SHAH | AFRICA | *Many theories purport to explain the purpose of zebra markings. Some contend that the patterns make for effective camouflage at water holes. Others hold that the stripes create an optical illusion designed to thwart predators from singling out a target.*

TIM LAMAN | BORNEO | *A giant pill millipede rolls up into a nearly impenetrable armored ball to protect itself, and unrolls when the coast is clear.*

DEADLY NEUROTOXIN

stored in the skin of the rough-skin newt has made it a disagreeable meal for most predators. Known as tetrodotoxin, the poison can kill within minutes of being ingested—and sometimes that is soon enough for the amphibian to outlive its would-be hunter: Newts have been observed crawling unharmed from the mouths of dead frogs that had rashly swallowed them whole just minutes earlier.

PAGE 98: JOEL SARTORE | CALIFORNIA | *The reeking-spray defense of the striped skunk is a misfortune that many dog lovers and canines—among them this wary bulldog—know all too well.*

PAGE 99: SHIN YOSHINO | AUSTRALIA | *As if possessing both reptile and mammal DNA weren't strange enough, the male duck-billed platypus also counts venomous ankle spurs as part of its arsenal of oddity.*

OPPOSITE: JOEL SARTORE | OREGON | *The deadly tetrodotoxin secreted by the rough-skin newt is the same type of poison found in puffer fish. It paralyzes nerve and muscle function, causing respiratory failure.*

OPPOSITE: MATTIAS KLUM | SOUTH AFRICA | *Meerkats employ sentinels to stand guard and alert the clan to approaching predators.*
BIANCA LAVIES | FLORIDA | *Startled, this armadillo launches reflexively straight into the air.*

THOMAS SCHMIDT | TANZANIA | *Gazelles stay constantly on the alert for predators. Such vigilance is key to ungulate defense—and survival.*

ROBERT SISSON | UNITED STATES | *A tiger swallowtail caterpillar resembles bird droppings in its first stage. Later, the caterpillar adopts green camouflage and develops the false eyespots on its tail that make it look like the head of a snake.*

OPPOSITE: DARLYNE MURAWSKI | INDONESIA | *An ovulid snail—just discernible by its antennae—gets the best of both worlds as it mimics and parasitizes its soft coral host.*

Out of this nettle,

DANGER,

we pluck this flower, safety.

–WILLIAM SHAKESPEARE

OPPOSITE: SAM ABELL | ALASKA | *A well-defended porcupine pauses on a riverbank. Porcupine quills are barbed, and tend to work their way deeper into flesh over time if not extracted.*

DISPLAYS OF

CHAPTER THREE

AGGRESSION

OPPOSITE: MICHAEL NICHOLS | GABON | *Young male forest elephants spar in Langoué Bai in a form of testing that often occurs between peers.*

THE ART OF RITUALIZED FIGHTING

Competition for resources—food, territory, even potential mates—can produce a pageantry of escalating threats by rivals eager to settle disputes or establish dominance.

Craggy, desolate, and snow-laden year round, the subantarctic landscape of South Georgia Island is no place for a honeymoon. Yet each September its rocky shores are overrun by a host of 8,000-pound Casanovas—also known as male southern elephant seals. They have come to assemble harems from among the female seals that will gather at island rookeries over the coming months to give birth and mate again before returning to the sea. Although these breeding grounds might hold hundreds, even thousands, of females each mating season, only one in three bulls will secure the right to reproduce—and with it the possibility of passing down his genes to future lineages. To do so, he must first establish himself as a dominant male, capable of protecting a harem and beating back lesser bulls intent on stealing "his" females.

Rather than wage bloody, costly battles with every adversary, beachmasters—the highest ranking bulls—employ an intricate system of vocalization, posturing, and threat displays known as ritualized fighting to intimidate subordinate bulls and run them off. Rival intruders are greeted with a roar unique to each male. Beachmasters, generally the largest of the bunch, have the lowest and loudest roar frequencies, which indicate their colossal size (and potentially superior fighting ability). If the interloper does not recognize the bull's rank by sound alone, the dominant male may aggressively approach the challenger, firmly displaying the trunk from which elephant seals derive their name. From here the face-off may escalate to chasing, then to shoving matches, and finally to violent and often injurious biting.

Like elephant seals, ibex have been known to butt heads over prospective mates—yet they do it literally. During the rut, to prevent breaking the horns that serve as their protective weaponry, ibex stage a sequence of less destructive moves—among them pushing, wrestling, locking horns, and butting heads from a low position—before resorting to outright head-to-head clashes. Territorial animals, such as wolves, can wield their dominion over thousands of acres, scrapping regularly with other packs to defend their bounds. And the contests can extend from between packs to within a pack itself: It's up to the alpha wolves to enforce the complex social structure of the pack, and this they do via facial expressions, growling, and biting. Eagles frequently pirate food from one another, using a variety of vocalizations and wing displays to bully opponents before charging in for the fight—or backing off if the show proves too menacing. And sometimes, merely wanting it more counts—between eagles, the victor in such contests is often the one that needs the food more desperately.

Back on South Georgia Island, actions speak as loudly as roars, and they make it clear why the seals invest more time in low-cost displays than they do in full-scale confrontations. The purpose of their visit to South Georgia is to mate, so time spent fighting is time wasted: It lowers each combatant's odds of reproductive success. Across the animal kingdom, from alligator to arachnid, ritualized fighting helps adversaries gauge the severity of threats made by competitors while avoiding the all-out combat over resources that can result in wasted energy, lifelong injuries, and even death.

PAGE 114: IAN NICHOLS | CONGO | *A 300-pound silverback western lowland gorilla announces his presence—and dominance—with chest beatings and a low, booming roar.*

PAGE 115: DIETMAR NILL | EUROPE | *Two greenfinches square off in a confrontation on the wing.*

ABOVE: CYRIL RUOSO | FRANCE | *These male stag beetles use their outsize mandibles to settle a mating dispute.*

OPPOSITE: JAMES L. STANFIELD | ISRAEL | *Nubian ibex lock horns in a sparring competition, a common sight during the rutting season.*

OPPOSITE: RAYMOND GEHMAN | UTAH | *Two Utah prairie dogs—painted with black identifying marks by researchers—leap into battle over both territory and females.*

MITSUAKI IWAGO | TANZANIA | *A black rhinoceros sometimes charges a perceived threat simply to investigate or intimidate it. If the threat offers serious harassment, the rhino may launch a dangerous, full-scale charge.*

WINNER BY A NOSE

The *kee-honks* and roars being made by this adult male proboscis monkey are meant to ward off a rival. If the racket doesn't dissuade the competition, the monkey may resort to leaping and impassioned branch shaking. Yet rarely does this peaceful primate resort to physical confrontation. Scientists suspect the nose acts as a resonance chamber, amplifying alarm calls to alert neighbors.

OPPOSITE: TIM LAMAN | BORNEO | *Male proboscis monkeys use vocalization for more than running off rivals; researchers think it's also how they communicate with nearby groups to ensure proper spacing between bands.*

PAGE 122: MITSUAKI IWAGO | TANZANIA | *Two male lions engage in a sanguinary battle. Males will sometimes form a coalition in a bid to seize control of a neighboring pride. The prize: a new harem of lionesses. The price: sometimes, their lives.*

PAGE 123: TIM LAMAN | JAPAN | *Going on the attack, a whooper swan chases after a rival in a skirmish over a female.*

KEENPRESS | NORWAY | *Two adult polar bears engage in a roaring match over the carcass of a minke whale. A cub observes from close range.*

The supreme art of

WAR

is to subdue the enemy
without fighting.

–SUN TZU

OPPOSITE: MICHAEL & PATRICIA FOGDEN | AUSTRALIA | *When startled, the frilled lizard fans out the large flap of skin around its neck in an effort to look larger.*

SERGEY GORSHKOV | RUSSIA | *Two Steller's sea eagles—an adult with white-topped wings and a juvenile—wage an angry scuffle over a fish. Though the sea eagle is well equipped with beak and talons to procure food of its own, it frequently pirates meals from other birds.*

WILD ENCOUNTERS

MELISSA FARLOW on photographing a heated stallion battle:

"I've never been afraid of horses," says photographer Melissa Farlow, *"though the only one I ever owned was a sad-looking, one-eyed pinto my parents gave me when I was six. I thought he was the most beautiful creature in the world."*

Wild horses were another matter. Guided by Karen Sussman, president of the International Society for the Protection of Mustangs and Burros on the Cheyenne River Sioux Indian reservation in South Dakota, Farlow learned how to approach wild horses, gain their trust, and even walk along as part of the herd.

Farlow's painstaking approach meant she was on the scene, camera in hand, when the herd's dominant stallion got into some violent confrontations with younger studs challenging him for his mares. "When wild stallions battle," she reports, "it's shockingly brutal. They go for each other's throats, they rip ears off, they bite and kick each other.

"Once you've heard the thud! thud! of hooves on chests up close," she reflects, *"it shakes you to the core. You'll never think of horses as tame and docile again."*

ABOVE: PHOTOGRAPHY BY FAITH | SOUTH DAKOTA | *Curious wild horses investigate photographer Melissa Farlow and her camera equipment.*

OPPOSITE: MELISSA FARLOW | SOUTH DAKOTA | *Two rearing stallions appear locked in a combative embrace as they battle for the right to mate with the mares of a wild herd.*

DANCE OFF!

Male Panamint rattlesnakes engage in a ritualized pattern known as a combat dance, marked by rhythmic head movements and often resulting in the two becoming intertwined as one attempts to overtake the other. This method of fighting is common to many types of snakes worldwide, history suggests: The caduceus—two snakes braided around Hermes's staff, a longtime symbol of the medical profession—may have been inspired by an ancient serpentine combat dance.

PAGE 132: MICHAEL NICHOLS | CONGO | *In a flurry of fur and flashing teeth, two gorillas clash over territory and dominance in the Léfini Faunal Reserve, a safe haven for orphaned gorillas.*

PAGE 133: GORDON GAHAN | AUSTRALIA | *Eastern gray male kangaroos use their muscular tails to support themselves as they box, periodically leaning back and kicking at each other with their hind legs.*

OPPOSITE: GORDON WILTSIE | CALIFORNIA | *For centuries, male snakes embroiled in combat dances like this one were mistakenly thought to be male-female pairs engaged in a mating ritual. (Sex in the limbless reptiles can be difficult to discern at a glance.)*

JOHN EASTCOTT & YVA MOMATIUK | SOUTH GEORGIA ISLAND | *Two beachmasters—harem holders and dominant bulls both—face off in a bloody duel for supremacy. The victor, at left, has just sent his rival packing.*

*One of the first businesses
of a sensible man is
to know when he is*

BEATEN

and to leave off fighting at once.

–SAMUEL BUTLER

MICHAEL NICHOLS | ETHIOPIA | *A male gelada monkey (opposite) opens its mouth and retracts its lips to display a pink expanse of skin and exceedingly large canines. Fearsome though those teeth may be, here female geladas rule the roost— as when three generations join in chiding an off-camera male (above, left). Males have little say in the lives of their family groups, but that didn't stop the one above at right from putting on a ferocious display toward a female who had strayed too close to a group of bachelors. Screeching, she calls for help to her girlfriends (who came to her rescue).*

OPPOSITE: MICHAEL NICHOLS | CENTRAL AFRICAN REPUBLIC | *A female forest elephant named Ladybeard by a researcher charges the scent of the photographer. In the background, Ladybeard's mother drapes a protective trunk over a calf.*

JIM & JAMIE DUTCHER | IDAHO | *A dominant wolf gives a muzzle bite to a lower-ranking pack member. Jaw sparring like this can be used to settle disputes or to reinforce pack order.*

TRUTH IN ADVERTISING

The roar of the red deer stag advertises—factually—both his size and his stamina. During rutting season, a lone stag will often approach a harem holder and initiate a ritualized roaring duel that can last for several minutes. If there is no clear winner, the two enter a "parallel walk," sizing each other up warily at close range. One stag may then lower his head to taunt the other into locking antlers. If accepted, they will separate and collide several times, rarely stopping until one gets injured or bolts.

OPPOSITE: CYRIL RUOSO | DENMARK | *When stags forsake the relative safety of ritualized fighting, they mean business: If a stag slips while locking antlers with another, his rival will often quickly spike the fallen one in a vulnerable area such as the eyes or front legs.*

THE PRIOR

ITY OF PLAY

LIFE—OR—DEATH LESSONS

For many animals, play may have a serious purpose: It fosters physical dexterity, problem-solving aptitudes, and flexible thinking—skills that may prove lifesaving in adulthood.

Some of the last wild chimpanzees in Africa reside on the forest-covered shores of Lake Tanganyika in western Tanzania. Here, in the remote and roadless Mahale Mountains National Park, lives a resourceful group of primates that scientists have studied for more than 40 years. The chimps are proficient at using tools and medicinal herbs. They also exhibit an enduring love of play.

As it does in many other animals, play among chimpanzees begins at a young age. Often it is initiated by an invitation—a gesture that may be as subtle as a stare or a relaxed, open-mouthed expression known as a play face. Or the invite may take the form of a more pronounced enticement—ground stamping, for example, branch shaking, or even a frisky slap or shove of a nearby companion. Once the invitation is accepted, play begins. It is marked by an exaggerated jumble of physical gestures, postures, and movements that mirror certain other aspects of chimpanzee life. The chasing, wrestling, hitting, and biting exchanged in a bout of play likewise occur in aggressive encounters between adult chimps. In the case of play, however, stronger, older chimps are noticeably gentle with youngsters: They refrain from making threatening displays, and they use "play markers"—facial expressions and vocalizations such as the "play pant," which sounds a bit like human laughter—to temper normally antagonistic actions. Young and old, chimps at play use these signals to send a message to one another—and to watchful mothers nearby, who have been known to break up play-fights that too closely resemble the real thing—that the rough-and-tumble mischief is all in good fun.

Play is not uniform across all animal species, and its genesis remains the subject of debate. One reason why animal play is so diverse may be that different types of play help animals of different lifestyles form a variety of necessary skills. Could the "locomotor play" of pronghorn fawns—characterized by swift running, twists, and leaps—serve to hone escape from future predators? Does the social play-fighting of coyote pups teach them combat tactics that will come in handy for aggressive encounters within the pack? Or might it begin to establish their rank within the group? When ravens engage in "object play"—raucous episodes of tug-of-war with twigs—does this help develop their innovative foraging techniques and their deft use of tools? Or do all these activities simply fine-tune cognitive abilities to prepare animals for one of the most commonly encountered scenarios in the natural realm: the unexpected?

Whatever the purpose of play, many animal behaviorists agree that it is an inherently cooperative enterprise; it's difficult, after all, to force someone to play. Many species—not just chimps—use play markers as unequivocal signals to prolong fun interactions, assuring players and bystanders alike that no breach of etiquette is intended. Behavioral researchers have interpreted this empirical evidence to mean that within complex communities like those of the Mahale chimps—hierarchical and highly political—the majority may adhere to certain social rules (a primary one: "Don't roughhouse with the baby"). This, paired with the fact that a sense of fair play among chimps seems to be the norm, is being further investigated by some scientists as possible evidence of social morality in our chimpanzee cousins, with whom we share 98 percent of our DNA.

MICHAEL NICHOLS | GABON | *Western lowland gorillas roughhouse in the safety of Gabon's Mpassa Reserve.*
Many of them were orphaned as a consequence of the bush-meat trade in central Africa.

PAGE 152: INGO ARNDT | GERMANY | *A pair of gray seals scrap lightheartedly in the shallows of the North Sea.*
While playing, juveniles will sometimes rise up and lunge at one another in mock battle; similar moves
are seen in aggressive interactions between adult males.

PAGE 153: JIM BRANDENBURG | MINNESOTA | *Timber wolf pups play with a partial skeleton in the Minnesota woods.*
Like chimpanzees, wolves use play markers—most frequently a deep bow—to solicit a good romp.

Play is a serious

BUSINESS.

–MARC BEKOFF

OPPOSITE: MICHAEL NICHOLS | INDIA | *Captive born, these two young Bengal tigers could not survive if released into the wild. Yet they still share much with their wild counterparts, including a love of play-fighting.*

PAGE 156: ROBERT CAPUTO | NEW HAMPSHIRE | *A pair of black bear siblings use their ears, facial expressions, and sometimes playful but painful biting to signal their receptiveness to play.*

PAGE 157: JIM BRANDENBURG | MINNESOTA | *Tug-of-war or life-skills class? This bout of object play between two ravens may inculcate cognitive skills that later prove crucial: Ravens have been observed driving off human intruders by dislodging overhead rocks that fall onto the invaders.*

NORBERT ROSING | CANADA | *Seemingly endless frolic marks the summer days of arctic fox kits, who scamper and play-fight their way through the season. In so doing, they gain critical strength and agility: Come autumn, the kits will leave the sanctuary of family life and begin the long search for food in the lean Arctic tundra.*

MICHAEL NICHOLS | KENYA | *In a playful pulling match, two baby African elephants test the strength of their trunks (which contain approximately 100,000 muscles).*

LOVE AT FIRST BITE?

Spotted hyenas are most often born as twins,
with their eyes open and their canine and incisor
teeth intact. Congenitally combative, they bite one
another in earnest within minutes of birth in their
natal dens. Over the next few weeks, however,
the cubs' behavior changes: Once a dominant sibling
emerges, the cubs start to trade outright aggression
for bouts of social play, sparring impishly with their
littermates or mother.

OPPOSITE: SUZI ESZTERHAS | KENYA | *Within a few weeks of their birth, infant spotted hyenas supplant their initial aggression with play and are transferred to a communal den, where their newfound play behavior may help integrate them into the larger group.*

WILD ENCOUNTERS

NORBERT ROSING, on photographing surprising play partners:

The German photographer Norbert Rosing can thank his lucky sunspots for some iconic images he captured of one species at play with another. In 1988 Rosing was drawn to Churchill, Manitoba, to photograph the northern lights. There he befriended Brian Ladoon, a breeder of Canadian Eskimo sled dogs, who told Rosing a polar bear in the area had been trying to play with his dogs.

After waiting for three weeks, camera at the ready, Rosing saw a bear approaching from a distance. Ladoon kept about 60 sled dogs outside, each chained to a stake. "As the dogs barked and strained at their tethers," Rosing recalls, "the bear crawled within range of the last dog in the line, gaining its trust. **All of a sudden the dog started to play with the polar bear—which lay down on its back and extended a paw like some sort of peace offering.** Next they touched noses, and soon the dog was jumping on the bear's belly and head."

ABOVE: COURTESY OF NORBERT ROSING | CANADA | *Photographer Norbert Rosing settles behind an ice-block blind, ready to shoot.*

OPPOSITE: NORBERT ROSING | CANADA | *For several seasons, a group of four polar bears—playmates of Ladoon's sled dogs—stayed in the general vicinity, protecting the dogs from "outside" bears that might attack.*

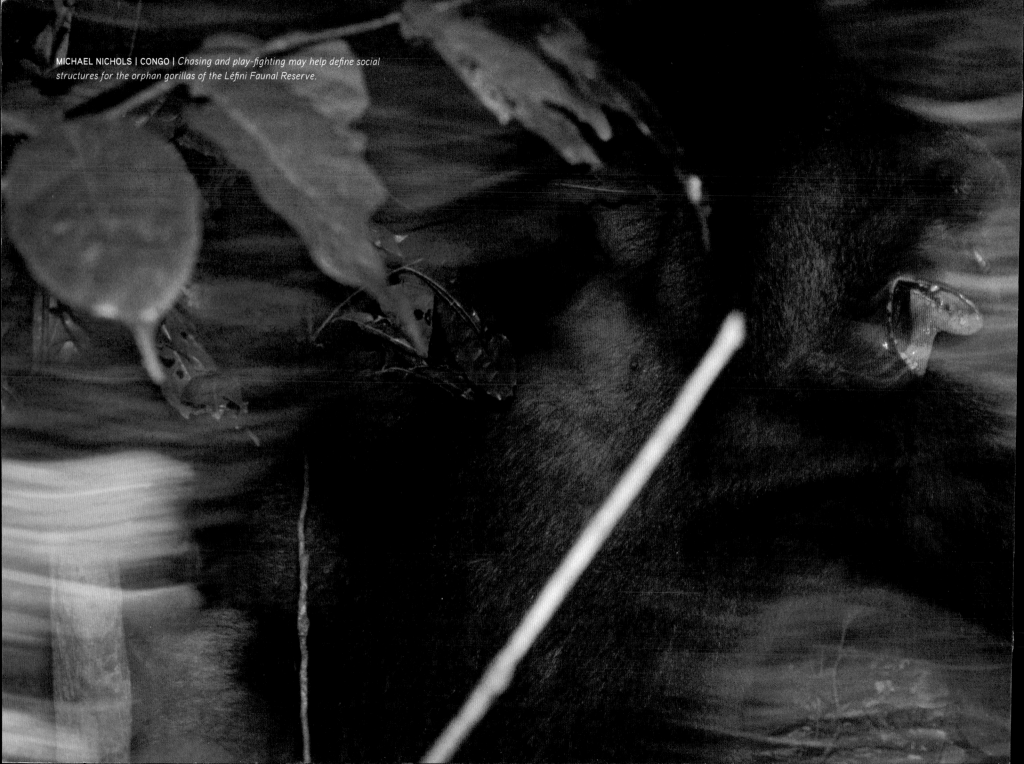

MICHAEL NICHOLS | CONGO | *Chasing and play-fighting may help define social structures for the orphan gorillas of the Léfini Faunal Reserve.*

*Those who will play with
cats must expect to be*

SCRATCHED.

–MIGUEL DE CERVANTES

OPPOSITE: CHRIS JOHNS | BOTSWANA | *Pouncing on his brother, a young African cheetah perfects his surprise attack. This playful behavior may help juvenile cheetahs sharpen their survival skills—training that is both timely and necessary, for soon they will be living on their own.*

JOEL SARTORE | MINNESOTA | *Members of the Ambassador Wolf Pack of the International Wolf Center bite and tussle in the snow. The center aims to educate the public about wolves, confident that as human appreciation of the species grows, so too will the wolf's chances of survival.*

CHRIS JOHNS | AFRICA | *Not just for dogs: A pair of young lions play a game of chase with a stick. Cubs will most often play when they feel secure, for instance when food is plentiful—and a lioness is standing vigil nearby.*

OPPOSITE: MATTIAS KLUM | SOUTH AFRICA | *At just one month old, these meerkat pups have grown enough to join group adults as they forage for food, but for now it's still playtime. The Kalahari Desert makes a dangerous home—only one in four pups will live to become an adult.*

MICHAEL NICHOLS | ETHIOPIA | *Frisky young gelada monkeys cavort amiably. Geladas live in large groups, and their play continues into adulthood. Primatologists conjecture that they use the interactions to strengthen social bonds.*

NO MONKEY BUSINESS

Bonobos have clear-cut signals when it comes to play. The most common sign is the "full play face," distinguished by an open mouth with the upper and lower teeth fully exposed. (This is sometimes also known as a "laughing" face.) Bonobos use this friendly expression to telegraph to one another mid-play that the exploit is meant to be fun, not aggressive.

PAGE 176: MELISSA FARLOW | KENTUCKY | *Foal play: A rare white Thoroughbred mare nibbles grass while her yearling gambols nearby. Most of the foal's strenuous exercise will come through play at this age.*

PAGE 177: STEVE WINTER | BRAZIL | *Able to kill some prey with a single bite to the head, a mother jaguar gives her cub a toothy but harmless play nip. The pair will stay together for approximately two years while she teaches the youngster to hunt.*

OPPOSITE: CYRIL RUOSO | DEMOCRATIC REPUBLIC OF THE CONGO | *Juvenile bonobo orphans play together in the Lola ya Bonobo Sanctuary. The name means "Paradise for Bonobos" in Lingala, a main language in nearby Kinshasa.*

{ ACKNOWLEDGMENTS }

Writing a book—even one with as few words as *Deadly Instinct*—takes a larger team than I ever imagined. Several people contributed ideas, expertise, and knowledge to this book, improving it vastly.

First, at National Geographic, I'd like to thank Barbara Brownell Grogan and Nina Hoffman for generously electing to pursue a concept on the basis of some rough layouts brought to them by a denizen of the art department; this—needless to say—is not how most books come to be. I'd add additional heartfelt thanks to Barbara for suggesting I write it; it was an incredible learning experience that I would never trade. Lisa Thomas, project manager and mentor extraordinaire, assembled the book team and held my hand through the process; if even a fraction of her grace and professionalism has rubbed off on me, I'll consider myself lucky. Maura Mulvihill, director of National Geographic's Image Collection, lent a contagious enthusiasm as well as countless great ideas; Debbie Li provided pivotal image research that set the bar high for the book's team. Photo editor Meredith Wilcox scoured the archive for unseen gems with her usual keen eye, and was unendingly patient and adept at fulfilling oddly specific requests (for toxic newts, lizards that shoot blood from their eyes, cannibalistic spiders, and so on). Jennifer Thornton and Judith Klein kept the team on track; they also taught me how to set up a bibliography and navigate the *Chicago Manual of Style* (sort of). Natalie Jones and Katie Juhl had fun, cool ideas to add to the mix that I very much appreciate. Finally, I'd like to thank Marianne Koszorus, the director of layout and design, for helping me juggle my projects so that I could focus on this one, and for providing important feedback on the layouts and structure of the book.

Lisa wisely chose to hire one of the best editors out there to polish and prune my amateur efforts: Allan "Conan the Grammarian" Fallow. His insight and flair made this a much better book and have made me sound much smarter than I actually am. His own writings—which appear in the "Wild Encounters" photographer interviews—are in fact my favorite parts of the book. I was lucky to have him on the project, as a friend, as an editor, and as a collaborator. Also: The man is punny.

Michelle Rae Harris, researcher, fact-checker, and all-around smarty-pants, went through the manuscript and captions with a fine-tooth comb and pulled out all the questionable bits and held them up to the light; she also assembled a crack team of experts who reviewed everything. Their feedback and fingerprints are all over this book, from images to captions to essays to literary quotes, and we could not have done it without them: Douglas Chadwick, wildlife biologist and author; John Paczkowski, Wildlife Conservation Society; Peter Fashing and Nga Nguyen, California State University Fullerton; Michael Douglas, University of Illinois; Kay E. Holekamp, Michigan State University; Atle Mysterud, University of Oslo, Norway; Tadahiro Murai, Kyoto University, Japan; Daniel R. Stahler, Yellowstone Wolf Project; Douglas Smith, Yellowstone Wolf Project; Edmund D. Brodie III, University of Virginia; P. J. N. de Bruyn, Mammal Research Institute, University of Pretoria; Andrea Turkalo, Wildlife Conservation Society; P. C. Lee, University of Stirling; David Houston, University of Glasgow; James Bradley, University of Bristol; George Watson, Smithsonian (Emeritus); William Brown, Fredonia University; Filippo Aureli, Liverpool John Moores University; Michael A. Huffman, Primate Research Institute, Kyoto University; Cristina Eisenberg, Oregon State University; Elisabetta Palagi, University of Pisa, Italy; Andrea Heydlauff, Managing Director, Panthera; Jess Edberg, International Wolf Center; and Mark Moffett. We also extend special thanks to the following experts who reviewed entire chapters: Jean M. L. Richardson, University of Victoria, British Columbia, Canada; Tagide deCarvalho, Carnegie Institution for Science; and Marc Bekoff, University of Colorado. Finally, our gratitude goes to Hugh Dingle of the University of California, Davis, who reviewed the book early on, answered many questions, and made invaluable recommendations.

National Geographic has a history of working with the best wildlife photographers in the world, and their collective efforts were the inspiration for this project. We are particularly grateful to six photographers—Melissa Farlow, Chris Johns, Darlyne Murawski, Norbert Rosing, Joel Sartore, and Brian Skerry—who took time out of their busy schedules to talk to us and share their stories and insights into the creatures they have encountered over the years, as well as their processes in the field. They were all such fascinating storytellers that their interviews became nearly impossible to edit.

Geographic also has a legacy of partnering with amazing scientists and explorers, among them Dr. Brady Barr, who contributed the foreword. His inspiring—and thrilling—work continues to educate, engage, and entertain, and we thank him for adding his voice, humor, and expertise to the project.

Finally, on a personal note, I'd like to thank my mother, my late father, and my sisters, for sitting through what I can only assume were a few too many stories about bugs, lizards, and predation; my friend Kristina Mazzocchi, who read through early drafts and told me which pictures she found too gross; my saintly and hilarious husband, Graham Caldwell, who patiently reviewed every page of the book (multiple times), told me what was boring, what was cool, and when I was finished; and my dog, Minnow, who—in addition to being adorable, loving, and smart—has taught me heaps about predation, aggression, play, and competition (particularly for food).

{ FURTHER READING }

PREDATORS & PARASITES

Alcock, John. *Animal Behavior: An Evolutionary Approach*. 8th ed. Sunderland, Mass.: Sinauer Associates, 2005.

Beschta, Robert L., and W. J. Ripple. "River Channel Dynamics Following Extirpation of Wolves in Northwestern Yellowstone National Park, USA." *Earth Surface Processes and Landforms* 31 (2006): 1525–1539.

Brown, Joel S., John W. Laundré, and Mahesh Garung. "The Ecology of Fear: Optimal Foraging, Game Theory, and Trophic Interactions." *Journal of Mammalogy* 80, no. 2 (1999): 385–399.

Creel, Scott, John Winnie, Jr., Bruce Maxwell, Ken Hamlin, and Michael Creel. "Elk Alter Habitat Selection as an Antipredator Response to Wolves." *Ecology* 86, no. 12 (2005): 3387–3397.

Howarth, F. G., S. H. Sohmer, and W. D. Duckworth. "Hawaiian Natural History and Conservation Efforts: What's Left Is Worth Saving." *BioScience* 38, no. 4 (1988): 232–237.

Laundré, John W., Lucina Hernández, and Kelly B. Altendorf. "Wolves, Elk, and Bison: Reestablishing the 'Landscape of Fear' in Yellowstone National Park, U.S.A." *Canadian Journal of Zoology* 79 (2001): 1401–1409.

Laundré, John W., Lucina Hernández, and William J. Ripple. "The Landscape of Fear: Ecological Implications of Being Afraid." *The Open Ecology Journal* 3 (2010): 1–7.

Lopez, Juan Carlos, and Diana Lopez. "Killer Whales (*Orcinus orca*) of Patagonia, and Their Behavior of Intentional Stranding While Hunting Near Shore." *Journal of Mammalogy* 66, no. 1 (1985): 181–183.

Manning, Adrian D., Iain J. Gordon, and William J. Ripple. "Restoring Landscapes of Fear With Wolves in the Scottish Highlands." *Biological Conservation* 142 (2009): 2314–2321.

Mao, Julie S., Mark S. Boyce, Douglas W. Smith, Francis J. Singer, David J. Vales, John M. Vore, and Evelyn H. Merrill. "Habitat Selection by Elk Before and After Wolf Reintroduction in Yellowstone National Park." *The Journal of Wildlife Management* 69, no. 4 (2005): 1691–1707.

Milton, Katharine. "Effects of Bot Fly (*Alouattamyia baeri*) Parasitism on a Free-Ranging Howler Monkey (*Alouatta palliata*) Population in Panama." *Journal of Zoology* 239 (1996): 39–63.

Moksnes, A., and E. Røskaft. "Adaptations of Meadow Pipits to Parasitism by the Common Cuckoo." *Behavioral Ecology and Sociobiology* 24, no. 1 (1989): 25–30.

Rilling, Susan, H. Mittelstaedt, and K. D. Roeder. "Prey Recognition in the Praying Mantis." *Behaviour* 14, no. 1/2 (1959): 164–184.

Ripple, William J., and Robert L. Beschta. "Linking Wolves and Plants: Aldo Leopold on Trophic Cascades." *BioScience* 55, no. 7 (2005): 613–621.

———. "Wolves and the Ecology of Fear: Can Predation Risk Structure Ecosystems?" *BioScience* 54, no. 8 (2004): 755–766.

Servedio, Maria R., and Russell Lande. "Coevolution of an Avian Host and Its Parasitic Cuckoo." *Evolution* 57, no. 5 (2003): 1164–1175.

Smith, Douglas W., Rolf O. Peterson, and Douglas B. Houston. "Yellowstone After Wolves." *BioScience* 53, no. 4 (2003): 330–340.

Voigt, Christian C., and Detlev H. Kelm. "Host Preference of the Common Vampire Bat (*Desmodus rotundus; Chiroptera*) Assessed by Stable Isotopes." *Journal of Mammalogy* 87, no. 1 (2006): 1–6.

ANIMAL DEFENSES

Blumstein, Daniel T. "Antipredatory Behavior: A Brief Overview." In *Encyclopedia of Animal Behavior*, edited by Marc Bekoff, 45–47. Westport, Conn.: Greenwood Press, 2004.

Brodie, Edmund D., Jr. "Investigations on the Skin Toxin of the Adult Rough-Skinned Newt, *Taricha granulosa*." *Copeia* 1968, no. 2 (1968): 307–313.

Brodie, Edmund D., III, and Edmund D. Brodie, Jr. "Costs of Exploiting Poisonous Prey: Evolutionary Trade-offs in a Predator-Prey Arms Race." *Evolution* 53, no. 2 (1999): 626–631.

———. "Predator-Prey Arms Races: Asymmetrical Selection on Predators and Prey May Be Reduced When Prey Are Dangerous." *BioScience* 49, no. 7 (1999): 557–568.

———. "Tetrodotoxin Resistance in Garter Snakes: An Evolutionary Response of Predators to Dangerous Prey." *Evolution* 44, no. 3 (1990): 651–659.

Dawkins, Richard. *The Greatest Show on Earth: The Evidence for Evolution*. New York: Free Press, 2010.

———. *The Selfish Gene*. 30th anniversary ed. Oxford: Oxford University Press, 2006.

Dawkins, Richard, and J. R. Krebs. "Arms Races Between and Within Species." *Proceedings of the Royal Society of London* 205, no. 1161 (1979): 489–511.

Soler, M., J. J. Soler, J. G. Martinez, and A. P. Moller. "Magpie Host Manipulation by Great Spotted Cuckoos: Evidence for an Avian Mafia?" *Evolution* 49, no. 4 (1995): 770–775.

Van Valen, Leigh. "A New Evolutionary Law." *Evolutionary Theory* 1 (1973): 1–30.

———. "The Red Queen." *The American Naturalist* 111, no. 980 (1977): 809–810.

Vermeij, Geerat J. "The Evolutionary Interaction Among Species: Selection, Escalation, and Coevolution." *Annual Review of Ecology and Systematics* 25 (1994): 219–236.

Williams, Becky L., Edmund D. Brodie, Jr., and Edmund D. Brodie III. "A Resistant Predator and Its Toxic Prey: Persistence of Newt Toxin Leads to Poisonous (Not Venomous) Snakes." *Journal of Chemical Ecology* 30, no. 10 (2004): 1901–1919.

ANIMAL COMPETITION

Alvarez, Fernando. "Horns and Fighting in Male Spanish Ibex, *Capra pyrenaica*." *Journal of Mammalogy* 71, no. 4 (1990): 608–616.

Archer, John. *The Behavioural Biology of Agression*. Cambridge: Cambridge University Press, 1988.

Caras, Roger A. *Dangerous to Man: The Definitive Story of Wildlife's Reputed Dangers*. 2nd ed. New York: Holt, Rinehart and Winston, 1975.

Clutton-Brock, T. H., and S. D. Albon. "The Logical Stag: Adaptive Aspects of Fighting in Red Deer *(Cervus elaphus)." Animal Behaviour* 27 (1979): 221–225.

———. "The Roaring of Red Deer and the Evolution of Honest Advertisement." *Behaviour* 69, no. 3/4 (1979): 145–170.

Endicott, Rachel. "Parents Desert Newborns After Break-in! Mother Goes for Food as Siblings Battle to Death! Father Eats Babies!" In *Encyclopedia of Animal Behavior*, edited by Marc Bekoff, 223–224. Westport, Conn.: Greenwood Press, 2004.

Fox, M. W. "A Comparative Study of the Development of Facial Expressions in Canids: Wolf, Coyote and Foxes." *Behaviour* 36, no. 1/2 (1970): 49–73.

Hansen, Andrew J. "Fighting Behavior in Bald Eagles: A Test of Game Theory." *Ecology* 67, no. 3 (1986): 787–797.

Johnstone, Rufus A. "Eavesdropping and Animal Conflict." *Proceedings of the National Academy of Sciences U.S.A.* 98, no. 16 (2001): 9177–9180.

Lundrigan, Barbara. "Morphology of Horns and Fighting Behavior in the Family Bovidae." *Journal of Mammalogy* 77, no. 2 (1996): 462–475.

McCann, T. S. "Aggression and Sexual Activity of Male Southern Elephant Seals, *Mirounga leonina." Journal of Zoology* 195 (1981): 295–310.

Smith, Douglas W. "Wolf Behavior: Learning to Live in Life or Death Situations." In *Encyclopedia of Animal Behavior*, edited by Marc Bekoff, 1181–1185. Westport, Conn.: Greenwood Press, 2004.

Smith, J. Maynard. "Game Theory and the Evolution of Behaviour." *Proceedings of the Royal Society of London* 205, no. 1161 (1979): 475–488.

———. "The Theory of Games and the Evolution of Animal Conflicts." *Journal of Theoretical Biology* 47 (1974): 209–221.

Watson, Paul, and Tagide deCarvalho. "Aggressive Behavior: Ritualized Fighting." In *Encyclopedia of Animal Behavior*, edited by Marc Bekoff, 7–12. Westport, Conn.: Greenwood Press, 2004.

ANIMAL PLAY

Bekoff, Marc. "Play: Social Play Behavior and Social Morality." In *Encyclopedia of Animal Behavior*, edited by Marc Bekoff, 833–845. Westport, Conn.: Greenwood Press, 2004.

———. "Play Signals as Punctuation: The Structure of Social Play in Canids." *Behaviour* 132, no. 5/6 (1995): 419–429.

———. "Social Play Behavior." *BioScience* 34, no. 4 (1984): 228–233.

———. "Social Play Behaviour: Cooperation, Fairness, Trust, and the Evolution of Morality." *Journal of Consciousness Studies* 8, no. 2 (2001): 81–90.

de Waal, Frans B. M. "The Communicative Repertoire of Captive Bonobos *(Pan paniscus)*, Compared to That of Chimpanzees." *Behaviour* 106, no. 3/4 (1988): 183–251.

Fagen, Robert. "Selective and Evolutionary Aspects of Animal Play." *The American Naturalist* 108, no. 964 (1974): 850–858.

Flack, Jessica C., Lisa A. Jeannotte, and Frans B. M. de Waal. "Play Signaling and the Perception of Social Rules by Juvenile Chimpanzees *(Pan troglodytes)." Journal of Comparative Psychology* 118, no. 2, (2004): 149–159.

Funk, Mildred Sears. "Play: Birds at Play." In *Encyclopedia of Animal Behavior*, edited by Marc Bekoff, 830–833. Westport, Conn.: Greenwood Press, 2004.

Hayaki, Hitoshige. "Social Play of Juvenile and Adolescent Chimpanzees in the Mahale Mountains National Park, Tanzania." *Primates* 26, no. 4 (1985): 343–360.

Matsusaka, Takahisa. "When Does Play Panting Occur During Social Play in Wild Chimpanzees?" *Primates* 45 (2004): 221–229.

Palagi, Elisabetta, and Tommaso Paoli. "Social Play in Bonobos: Not Only an Immature Matter." In *Bonobos: Behavior, Ecology, and Conservation*, by Takeshi Furuichi and Jo Myers Thompson, 55–74. New York: Springer-Verlag, 2008.

Pollick, Amy S., and Frans B. M. de Waal. "Ape Gestures and Language Evolution." *Proceedings of the National Academy of Sciences U.S.A.* 104, no. 19 (2007): 8184–8189.

Spinka, Marek, Ruth C. Newberry, and Marc Bekoff. "Mammalian Play: Training for the Unexpected." *The Quarterly Review of Biology* 76, No. 2 (2001): 141–168.

Wilson, Susan C., and Devra G. Kleiman. "Eliciting Play: A Comparative Study." *American Zoologist* 14, no. 1 (1974): 341–370.

ADDITIONAL RESOURCES:

All about Animals:
http://animals.nationalgeographic.com/animals/
http://animaldiversity.ummz.umich.edu/site/index.html

Animals in Yellowstone National Park:
www.nps.gov/yell/naturescience/animals.htm

Facts about Primates:
http://pin.primate.wisc.edu/
http://primatecenter.duke.edu/
www.greatapetrust.org/

Animal Organizations:
www.defenders.org/index.php
www.awionline.org/
www.wolf.org/wolves/index.asp

{ ANIMAL INDEX }

DEADLY INSTINCT

Melissa Farris

PUBLISHED BY THE NATIONAL GEOGRAPHIC SOCIETY

John M. Fahey, Jr., *President and Chief Executive Officer*
Gilbert M. Grosvenor, *Chairman of the Board*
Tim T. Kelly, *President, Global Media Group*
John Q. Griffin, *Executive Vice President; President, Publishing*
Nina D. Hoffman, *Executive Vice President;*
  President, Book Publishing Group

PREPARED BY THE BOOK DIVISION

Barbara Brownell Grogan, *Vice President and Editor in Chief*
Marianne R. Koszorus, *Director of Design*
Carl Mehler, *Director of Maps*
R. Gary Colbert, *Production Director*
Jennifer A. Thornton, *Managing Editor*
Meredith C. Wilcox, *Administrative Director, Illustrations*
Lisa Thomas, *Senior Editor*

STAFF FOR THIS BOOK

Melissa Farris, *Project Editor, Art Director*
Meredith Wilcox, *Illustrations Editor*
Allan Fallow, *Text Editor; author, Wild Encounter interviews*
Michelle Rae Harris, *Researcher*
Debbie Li, *Illustrations Research*
Judith Klein, *Production Editor*
Mike Horenstein, *Production Manager*

MANUFACTURING AND QUALITY MANAGEMENT

Christopher A. Liedel, *Chief Financial Officer*
Phillip L. Schlosser, *Senior Vice President*
Chris Brown, *Technical Director*
Nicole Elliott, *Manager*
Rachel Faulise, *Manager*
Robert L. Barr, *Manager*

THE NATIONAL GEOGRAPHIC SOCIETY is one of the world's largest nonprofit scientific and educational organizations. Founded in 1888 to "increase and diffuse geographic knowledge," the Society works to inspire people to care about the planet. It reaches more than 325 million people worldwide each month through its official journal, *National Geographic*, and other magazines; National Geographic Channel; television documentaries; music; radio; films; books; DVDs; maps; exhibitions; school publishing programs; interactive media; and merchandise. National Geographic has funded more than 9,000 scientific research, conservation, and exploration projects and supports an education program combating geographic illiteracy. For more information, visit nationalgeographic.com.

For more information, please call 1-800-NGS LINE (647-5463) or write to the following address:

National Geographic Society
1145 17th Street N.W.
Washington, D.C. 20036-4688 U.S.A.
Visit us online at www.nationalgeographic.com

For information about special discounts for bulk purchases, please contact National Geographic Books Special Sales: ngspecsales@ngs.org

For rights or permissions inquiries, please contact National Geographic Books Subsidiary Rights: ngbookrights@ngs.org

ISBN: 978-1-4262-0725-9

Printed in China
11/RRDS/1

ADDITIONAL ILLUSTRATIONS CREDITS:
Animals Animals/Earth Scenes: 78

Minden Pictures: 76 (right), 157

Minden Pictures/National Geographic Stock: Cover, 1, 8-9, 22, 31, 34, 47, 48-49, 61, 67 (FLPA), 68, 69, 70, 74, 84-85, 89, 93, 99, 115 (Foto Natura), 116 (both; JH Editorial), 119, 122, 127, 128-129 (all), 145 (JH Editorial), 152, 153, 162, 179 (JH Editorial), Back cover (center)

National Geographic My Shot: 26, 76 (left), 104-105

National Geographic Stock: 12-13, 58-59, 86, 91, 96-97 (all), 98, 101, 107, 124-125, 134, 136-137, 142, 143, 164, 165

National Geographic Television: 14

National Park Service: 10-11